The Best of Bakelite
& other plastic jewelry

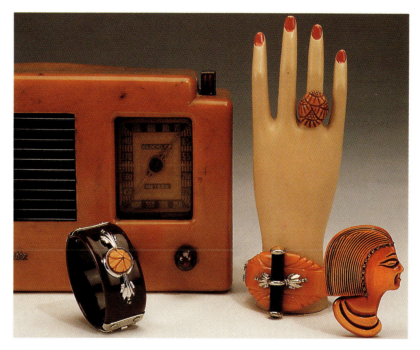

by Dee Battle & Alayne Lesser

Photography by Doug Congdon-Martin

Schiffer Publishing Ltd

77 Lower Valley Road, Atglen, PA 19310

Dedication

To Dee's best friend
and partner in business and life,
her husband Bill
&
to Joshua and Mischele Lesser.

title page:

Two superior German metalwork and Bakelite hinged bracelets, a super Bakelite radio by Lafayette, carved ring and Egyptian head brooch.

Copyright © 1996 by Dee Battle and Alayne Lesser.
Library of Congress Catalog Card Number: 95-74955

Printed in Hong Kong.
ISBN: 0-88740-901-6

Published by Schiffer Publishing, Ltd.
77 Lower Valley Road
Atglen, PA 19310
Please write for a free catalog.
This book may be purchased from the publisher.
Please include $2.95 postage.
Try your bookstore first.

We are interested in hearing from authors
with book ideas on related subjects.

The Bakelite Logo

The registered trademark for Bakelite was a capital "B" enclosed in a trefoil. Below the "B" appeared the mathematical sign for infinity, perhaps indicating the number of uses for Bakelite products—for they were numerous and varied.

Contents

Acknowledgments

Dee and Alayne would like to give special thanks to Dee's husband Bill for his patience and understanding throughout this project, and to Alayne's son and daughter, Joshua and Mischele, for their encouragement. A special thanks goes to Jerry Fraley and Buddy Makin for bringing their wives (Barbara and Alice) to Atlanta for our photo session. Our deep appreciation to Charliene Felts for not only sharing her wonderful collection of Bakelite but also for providing us with great references on the history of Bakelite. This book could not have been written without the help of Nancy Schiffer—a millon thanks—and Doug Congdon-Martin for his great photos. We are truly greatful to the following people who graciously allowed us to photograph the great examples of Bakelite contained in this book:

Charliene S. Felts, collector and dealer at Creative Collections, 527 South Pineapple Ave., Sarasota, FL 34236. (813) 951-0477.

Barbara Fraley, private collector, Big Stone Gap, VA.

Alice Makin, private collector, Big Stone Gap, VA.

Sandy Crouse, collector and dealer at Patty & Friends, 1225 Ninth Street North, St. Petersburg, FL 33701. (813)-821-2106.

Julie Bechko, private collector, Sarasota, FL.

Janis Collier, private collector, Sarasota, FL.

Reva Levy, private collector, Sarasota, FL.

Adrian Reiff, collector and artist, Bradenton, FL.

Audrey Lewis, private colleector, Sarasota, FL

Plastics

The time has come, the chemist said,
to talk of many things;
Of poker chips, umbrella tips,
And combs and teething rings,
Of minnow traps and bottlecaps
And toys and botton-hooks,
Of banjo picks and magic tricks
And clasps and pocketbooks,
Of blotter tops, electric clocks,
The number is fantastic;
And all these things the chemist brings
Into the world with plastic.

—Edward Mabley, 1941

As Bakelite jewelry collecting continues its revival, and so many of us are still bewitched by the dazzling colors, intricate designs and configurations, one might say as we search high and low, "where have all the exquisite Bakelite novelties gone?" Long time hiding? That's probably no longer true. Ten serious Bakelite collectors and dealers from the Southeast United States are sharing their Best of Bakelite creations with you readers. We have assembled this book for the greatest display of Bakelite ever seen. Our purpose is to share knowledge and incredible photos to appeal to the advanced and novice jewelry lovers alike, and all people fascinated with design. The infatuation with Bakelite will never end...let the hunt begin!

Bakelite and Early Plastics History

Before Bakelite was available, the synthetic material cellulose acetate (patented as "celluloid" in 1867 by John Wesley Hyatt of Albany, New York) was made into a number of decorative products, including jewelry. Light in weight and easily molded into shapes with intricate designs, celluloid jewelry copied the styles of carved ivory. Several very fine celluloid jewelry articles are presented in this book, and their refinement reflects the lacy styles popular in the late nineteenth century. In the first decades of the twentieth century, French-made celluloid jewelry in pastel colors were imported to the United States.

Dr. Leo Hendrick Bakeland (1863-1944) is celebrated today as the father of the modern plastics industry. He started research in the study of phenolic resins in Belgium as early as 1885, but didn't pursue them further until after a career making photographic plates. His development of Velox photographic paper is well documented. In 1899, as an independently wealthy man in Yonkers, N.Y., Bakelanad resumed his laboratory studies of phenolic resins and worked with various chemicals to produce a thermoset resin that could not be melted by applications of heat after it hardened. His new material was literally "set for life;" it was christened "Bakelite," and on July 13, 1907, Dr. Bakeland obtained a patent for it. The demand

snapshot of factory

was so great for Bakelite in the industrial world that he formed the General Bakelite Company, in Perth Amboy, New Jersey. Early advertising called Bakelite "a material of a thousand uses." Some of those uses included electrical insulators and parts for the textiles and transportation industries, building and household goods, games, toys, and, of course, the wonderful jewelry pieces we search for today.

Phenolic resins (Bakelite) can be molded, and so cast-molded products were made by compressing powdered or granular compounds under heat. Examples of molded phenolics from the 1930s include telephones, radio cabinets, and industrial items.

Cast products also were made by pouring liquid resin into molds—usually made of lead, and letting them harden. When removed from the mold, the product was often hand carved or jig sawed before receiving its final polish. Examples of cast phenolics of the 1930s are small boxes, buttons, buckles, and cutlery handles. Most cast phenolic production ceased in 1942. After the

Second World War, most companies had switched to the manufacture of molded compounds.

James W. Munroe, a long-time employee of the Bakelite Corporation, whose father also worked with the company, has given us a wonderful testimony and recollection of Bakelite items. Mr. Munroe remembers:

The Bakelite Company relocated from Perth Amboy, N. J. to Bound Brook, N. J. in 1931. It is said that Dr. Bakeland established this plant to insure the employment of his many faithful employees from around the country.

The company sponsored many activities for the employees, such as skeet shooting, baseball (hard and soft), horseshoe pitching, tennis, golf, boat excursions up the Hudson River, Christmas parties, bridge clubs, bowling, and others.

I recall that at one time my family had seven (7) members employed at the plant.

The plant's test mold department processed various resins and molding materials for quality and durability by molding ash trays, discs, cutlery handles, automobile distributor caps, etc. In our house possibly twice a year or as needed, my father would acquire test pieces of cutlery handles and presto, we would have cutlery with green handles, next yellow, then brown, and so, always in style.

We would have various sizes of ash trays for the smokers- resin sticks which were heart shaped from which the girls could have heart pendants.

The disks were approximatelly two (2) inches in diamemter, various colors. Stacked tops to bottoms and placed on a hard wood base, they made a nice table lamp.

In 1939, the Bakelite Company displayed an all plastic airplane.

My forty-three (43) years at the plant is equalled by many others who say it was a great place to work.

Today, the commonly acceptable term "Bakelite" is used for all cast phenolics even though the Catalin Corpoation in 1935 manufactured 70% of the items we know as "Bakelite" today. The term "Bakelite" will be used throughout this book to identify phenolic as opposed to other plastics.

Phenolic resins are not light fast— they have changed color over the years. Many colors were available originally, but the colors have changed with time. Some of the original colors included purple, blue, pink, turquoise in addition to the common yellow, amber, orange, red and green. Because of the chemicals used in the formula, the items could darken or now may have a yellow cast, according to the amount and strength of the light the piece was subjected to. Bakelite has been unearthed after twenty years and still retained its lustre. In about 1930, white, ivory, and many pastels became available and they were considered light-fast. In 1935 there were foreign affiliations of the Bakelite Company making Bakelite in England, Italy, Germany, Japan and Canada. From each country came a variety of products including costume jewelry.

Bakelite articles were sold in the 1930s at United States department stores including Sears & Roebuck, Saks Fifth Avenue, B. Altman's, and Bonwit Teller, to name a few. Bakelite jewelry sold for prices ranging from a few cents per piece to several dollars for better pieces. French designer René Lalique even designed items using the material. It seems that wearing plastic jewely cuts across all socio-economic lines.

Bakelite Review, vol. 7, no. 3, Silver Anniversary, 1919-1935 with dynamic five-color hinged bracelets and buckle.

Research tells us there was never a clear-white Bakelite. A clear synthetic plastic product sold under the name Prystal was manufactured first in Germany and later in the United States in the early decades of the twentieth century. Prystal very likely was an acrylic and not a phenolic resin. When you see or think of clear acrylic plastics today, you think of Lucite and Plexiglas (these names are both trademarks) made from acrylic thermoplastic resins which were first produced in the United Staes in the mid-1930s. Several fine acrylic jewelry and accessory items (handbags, for example) are shown in this book, and many pieces combine Bakelite with acrylic details for variety.

The word "plastic" has sometimes been associated with products that are unreal, expressionless, devoid of any artistic merit or individuality. This unfortunate association obscures a very amusing and stylish range of collector's items that are presented here. Just a quick look at some of the jewelry and accessories in this book should put all the expression, life and individuality back into the perception of plastics.

Fun, Funky & Fantasy

A grouping of exquisite carved and metal inset bracelets and brooches in black and red. Red and black stretch bracelet and matching earrings—Lucite purse.

Bracelets

A group of asymmetrical Bakelite jewelry.

Beat'l ware bowl full of Bakelite bangles with dots, carved, laminated decorations.

Bangles of Bakelite with "gum drops" or bowtie decorations.

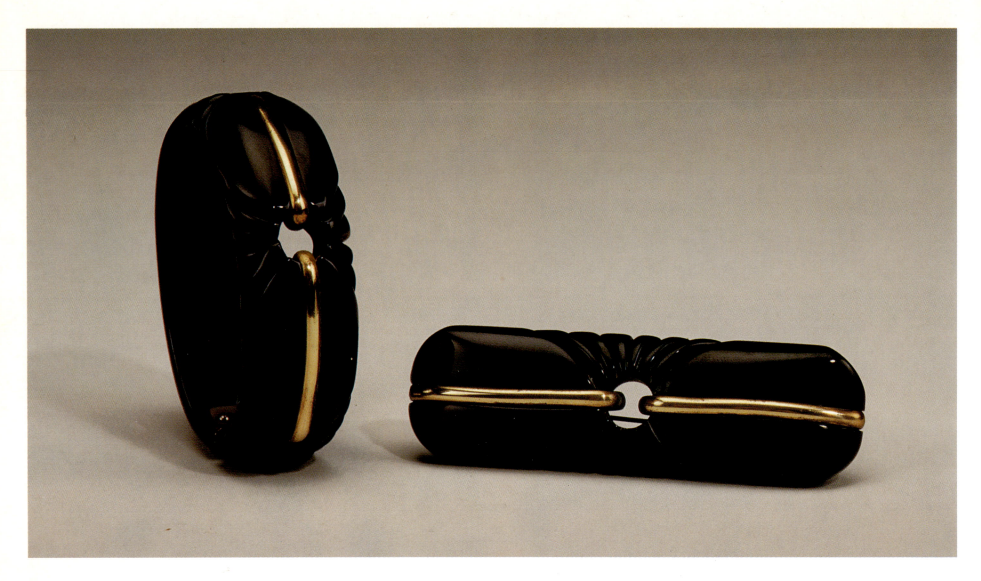

Black hinged bracelet and matching brooch.

Just Bakelite, just wonderful! including an opera bracelet with seven concealed compartments for cosmetics and mirrors—Horse mane of sterling silver—Lucite purse.

Bangles and hinged bracelets of Bakelite with beautiful carved designs.

Bakelite of distinction.

A grouping of Bakelite—clear, undercarved and layered jewelry—Lucite purse.

Bangles and hinged, clear Bakelite with undercarved designs, many with painted floral details.

Laminated, four-color carved bracelets.

Laminated Bakelite bangles and cuff.

German hinged bracelet.

*Polkadot decorated Bakelite bangle brace-
lets—some clear undercarved, or carved from
two layers to form dots.*

Layered Bakelite bangles with designs cut
through to the contrasting layer. Two of a kind
and three of a kind are amazing together.
The radio is part of the good news.

Bold Bakelite bangles with cut designs.

Red and yellow carved bracelet.

Red carved bangle bracelet.

Butterscotch Bakelite carved bangle bracelet.

Clear, hinged Bakelite bracelet with carved inset.

"Killer" cat and insect undercarved clear Bakelite bracelet.

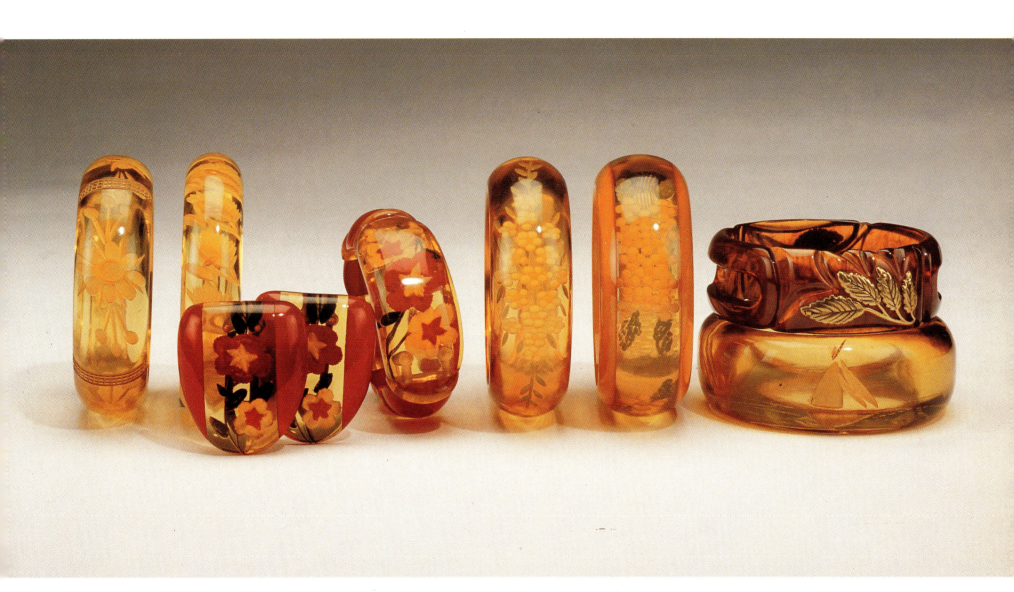

Outstanding undercarved transparent Bakelite bracelets and dress clips.

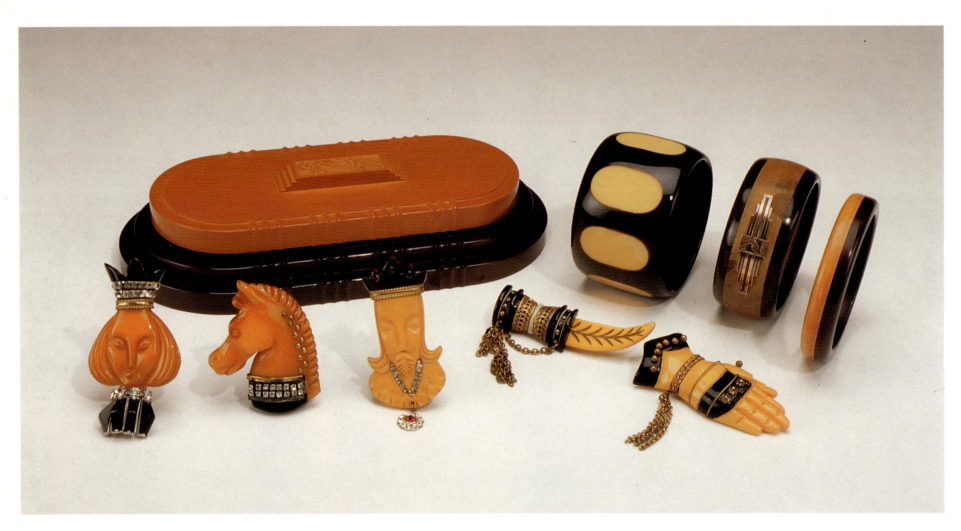

Brown tones of Bakelite in exquisite designs, to
include brooches as chess pieces. Hand and
sword depicting the Crusades era.

Chili peppers hinged bracelet.

Black and red dot bangle bracelet.

Carved Bakelite bangle bracelets.

Deeply carved bangle bracelet.

Floral carved bangle bracelet.

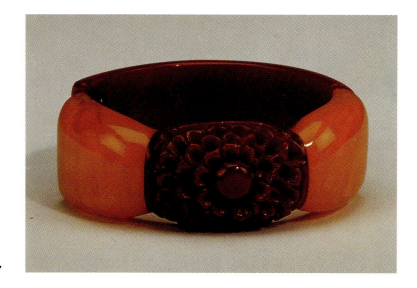

Two-toned carved hinged bracelet.

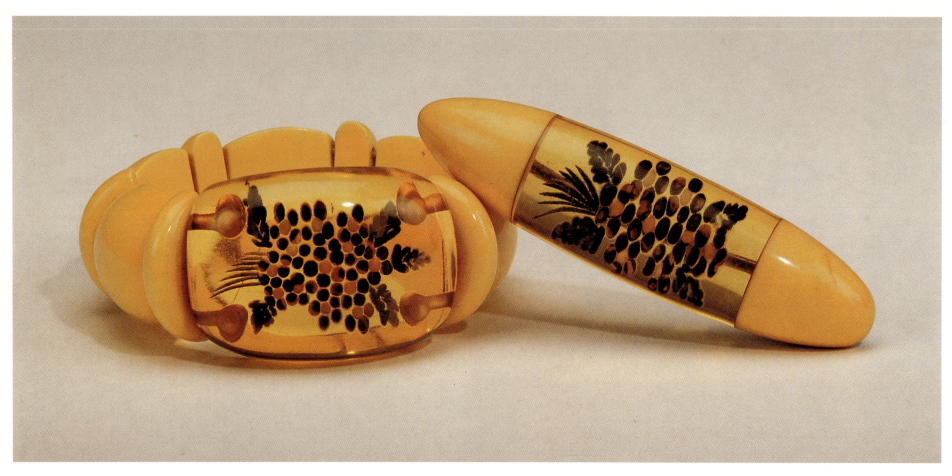

Undercarved and painted floral set.

Celluloid bangle bracelet with rhinestones.

Brown Bakelite lazy susan with a spectacular set of clear golden Bakelite with rhinestones.

Scotty bracelet.

Great water critters attached to hinged bracelets.

Gorgeous green, carved, hinged bracelets with
wonderful creatures and other designs - and a
Lucite purse.

Chic, unique, multi color cuff Bakelite bracelets

Necklaces

Creamy, dreamy butterscotch Bakelite grouping.

*Red and ready Bakelite—unusual necklace
with metal insets. Marvelous carved flower—
bangle with metal studs and poodles.*

Bakelite "gum drop" jewelry and Lucite purse.

Bakelite carved cameo bracelet with metal and Bakelite linked necklace or belt and earrings from Australia. Lucite purse.

Red passion

Carved ivory and amber simulated in Bakelite.

Ekco radio, necklace, bracelets, pair of bangles with earrings, and figural brooches with wood—all Bakelite.

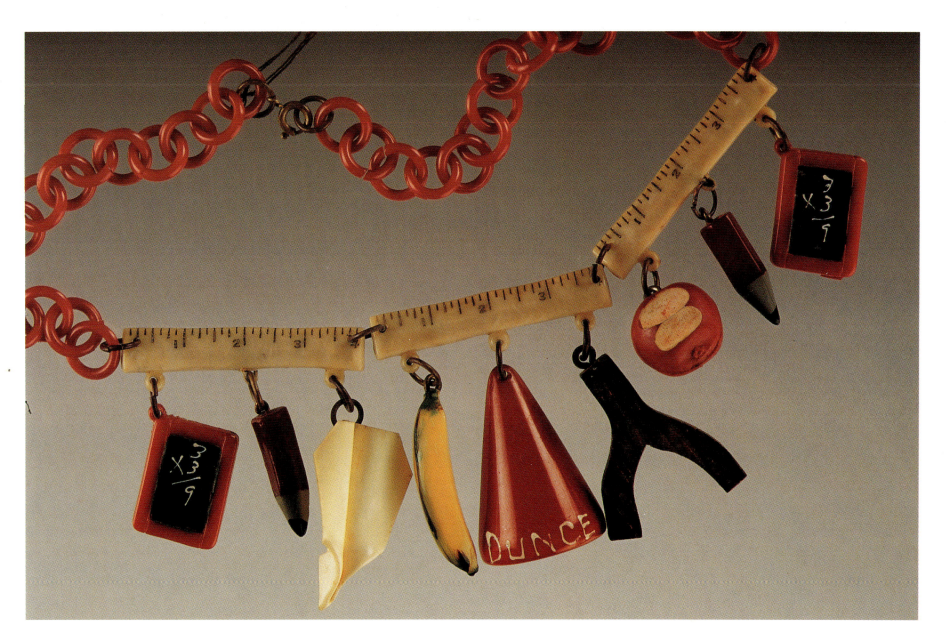

Bakelite necklaces, matching bracelet, and earrings—layered white and red bangle, compact with Lucite purse.

Link necklace with school days charms

Bow and scent bottle brooch, end of day Bakelite.

Lucite scent bottles.

Bakelite scent bottles.

Yardley perfume display.

Bakelite prototype perfume bottle—Amelia Earhart "trophy" and Airplane scent bottle.

Bakelite scent bottles .

Plastic scent bottles including a Toby figure and swinging door.

Bakelite scent bottles.

*Vanity boxes for powder, cards, and rouge,
made from Bakelite and Celluloid.*

Bakelite boxes for every useful purpose.

Scottie powder box.

A group of Bakelite and celluloid ring boxes.

Ring holder.

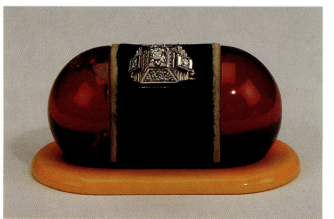

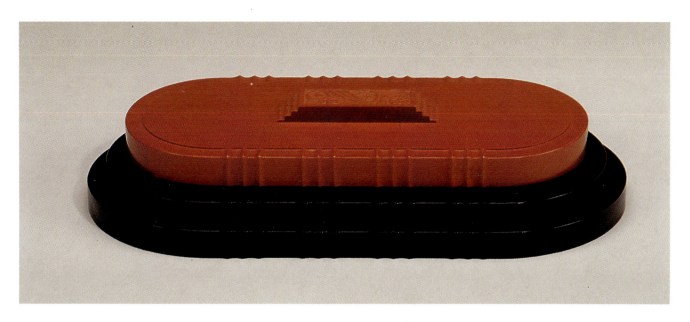

Nice two tone oval box.

Two tone oval box shown open

Smoking accessories in Bakelite: full-sized humidores, lighter, cigarette boxes, cigarette holder and ash tray.

Lighter.

o 64 o

Ring boxes in Bakelite.

Large Bakelite container with carved fish attached. Bakelite poker chips, three butter pat molds, and the Rolmonica which plays paper music rolls and was patented 11-3-25.

Watches

Bakelite wrist watches.

It's a good time for Bakelite.

Compacts

Make-up set.

For the dresser top, a Bakelite counter display for Tattoo lipstick samples, and four spectacular compacts.

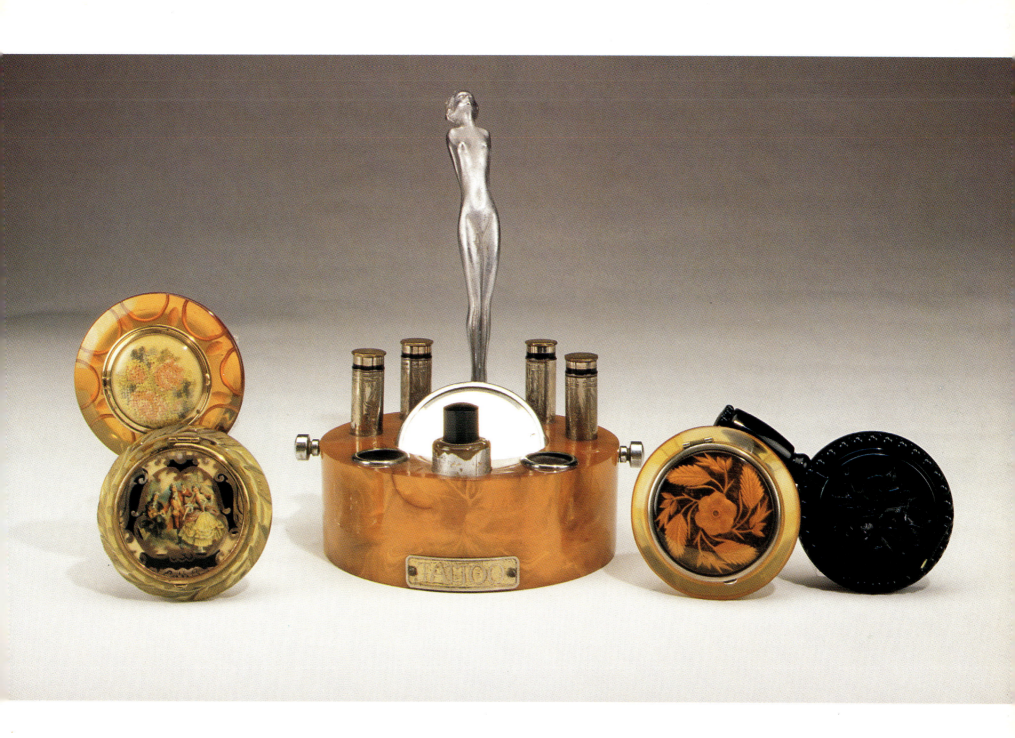

Carved compact purses from the early decades.

Celluloid card case, closed.

Celluloid card case, opened.

Black Bakelite evening bag and compact
purses.

Bakelite tortoise color purse.

The same tortoise color purse, open to show interior compartments .

Brooches

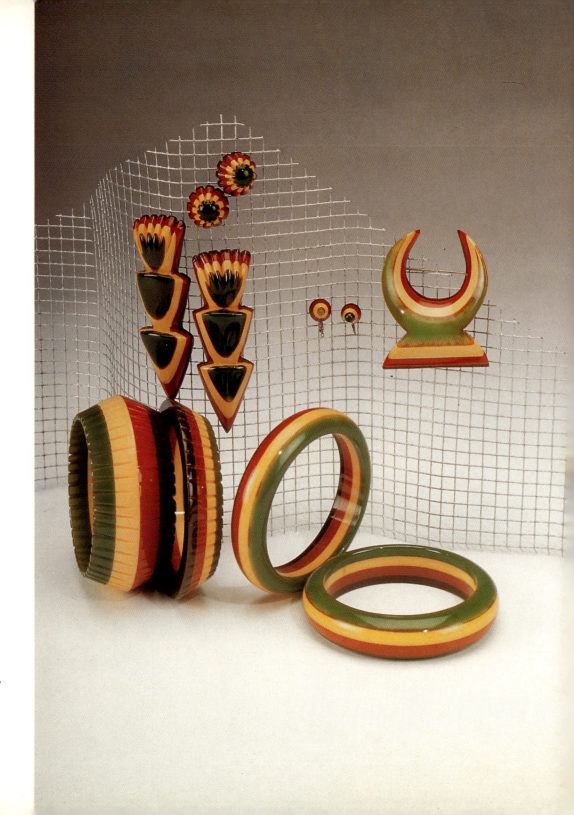

Matching set of laminated, carved, multi-color Bakelite jewelry including four bracelets, three brooches, and two pair of earrings. Very rare.

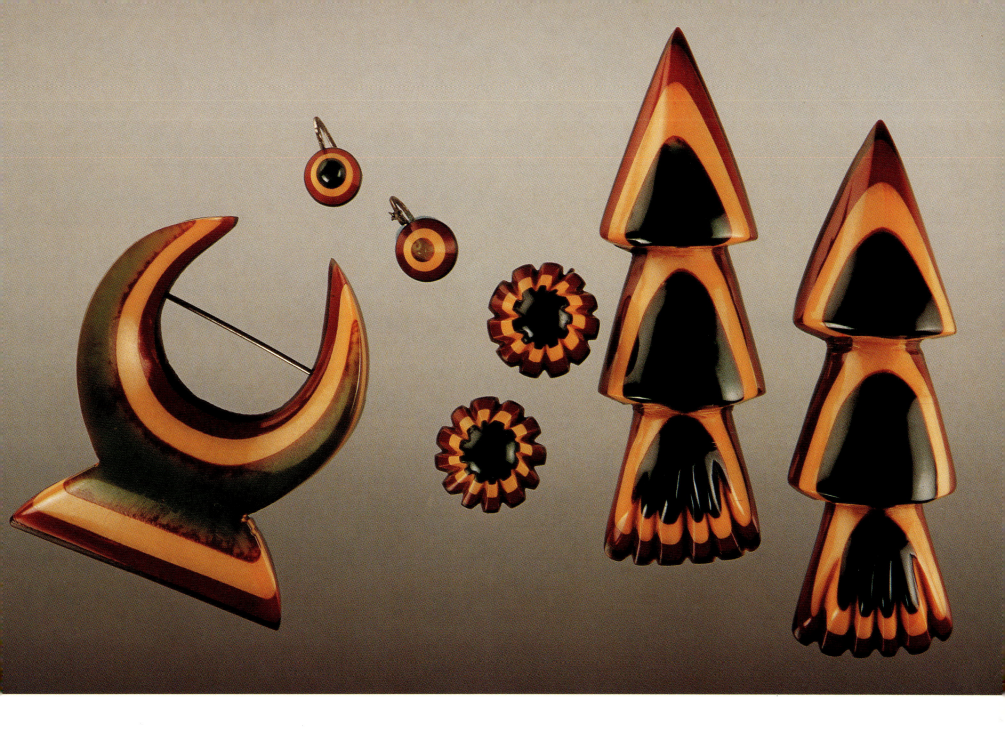

Four-color laminated matching set.

Bakelite school day memories.

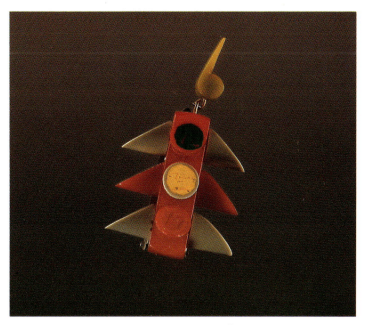

Traffic light charm.

School days brooch.

Tools brooch.

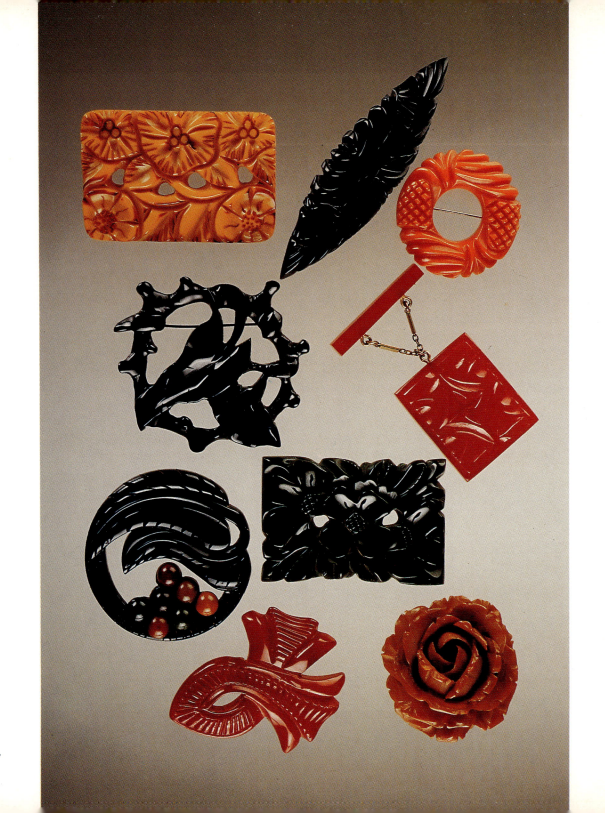

Carved Bakelite brooches.

Floral carved brooches.

Floral undercarved brooches.

Carved flowers.

South of the border
'Fiesta'.

Colorful earrings to add zip to the wardrobe.

Bakelite sets of carved, layered bracelets and earrings, layered clear oversize brooch on Lucite purse.

Discs of Bakelite, carved, layered, and stud-
ded with rhinestones, all became earrings—for
the fun of it.

Buttons

Buttons and rings of Bakelite and Lucite displayed in Lucite jewelry box.

Is your Bakelite showing?.

Figural novelty buttons.

Sew it on.

Buttons of all kinds.

Buttons by the dozen.

A day at the (Bakelite) beach with Lucite pail.

By the sea—wonderful bracelets and brooches of various colors and designs with metal studs, layered and carved.

Sail boat brooches.

Navy blue and nauticals, too, all Bakelite—
Lucite purse .

Patriotic

Red, White, and Blue Bakelite with Lucite purse.

Dewald radio in Bakelite case brings wartime news from the Front. The jewelry is also Bakelite.

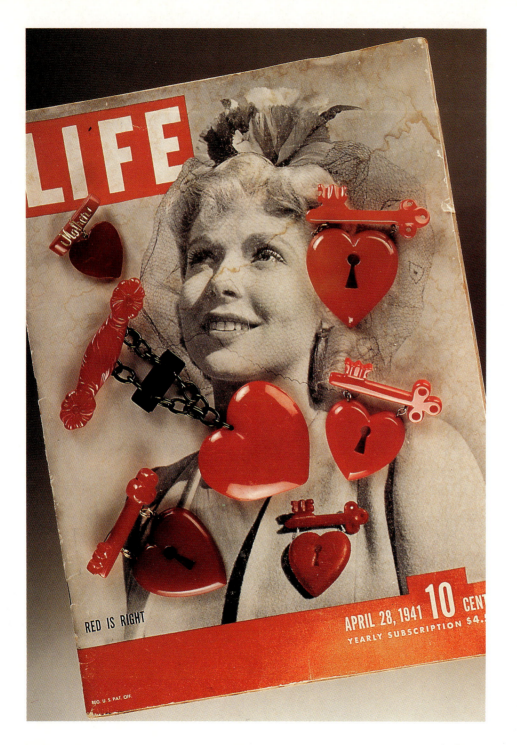

Heart of Hearts and cover of Life magazine,
April 28, 1941, featuring Bakelite key and
heart brooch.

Victory wings.

Sweetheart wings.

U. S. Navy sweetheart necklace.

Sweetheart jewelry.

Tablewares

Bakelite napkin holders.

Bakelite salt and pepper shakers and hors d'oeuvre picks.

Knives, forks and spoons, napkin rings, and salt and pepper shakers.

Salt and pepper shakers and flatware.

Serving tray—Bakelite from Australia—
great color.

Shake it up, Baby.

Shake and Shout.

Hats, Hands, Faces & Accessories

Bakelite hand brooch attributed to designer
Coco Chanel, with adjustable back for multiple
positions. 14 K gold plated ring and cuff.

Faces, hats and a hand.

Hats, hand and shoes brooches.

Hats, necklace, shoes, and hand Bakelite brooches.

Happening hats and fabulous faces.

Crib toys of times past.

Great carved and layered buckles.

Belt buckles.

Fruits & Veggies

Cherries Jubilee, in Bakelite!

Bakelite fruit set: carved oranges.

Fruit salad in Bakelite.

A Bakelite vegetable garden.

Bakelite bananas and carved fruit necklaces.

Chicquita Bananas?

Sports

Touchdown.

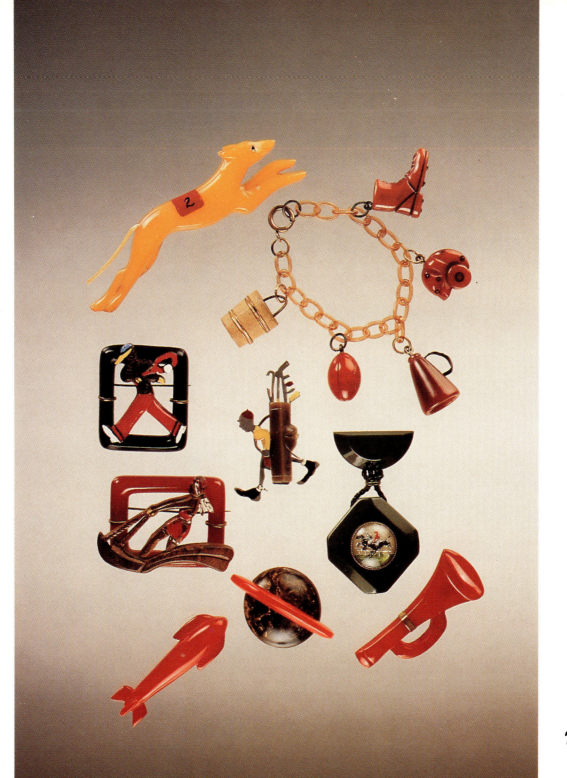

Wide world of sports.

Fidos and Feline.

Scotties galore. Bakelite bangle with metal scotties—Lucite purse.

A group of unusual Bakelite pieces—brooches
and bracelets, magnificent undercarved cat in
bangle, and large bangle with pewter butterfly.
Lucite purse.

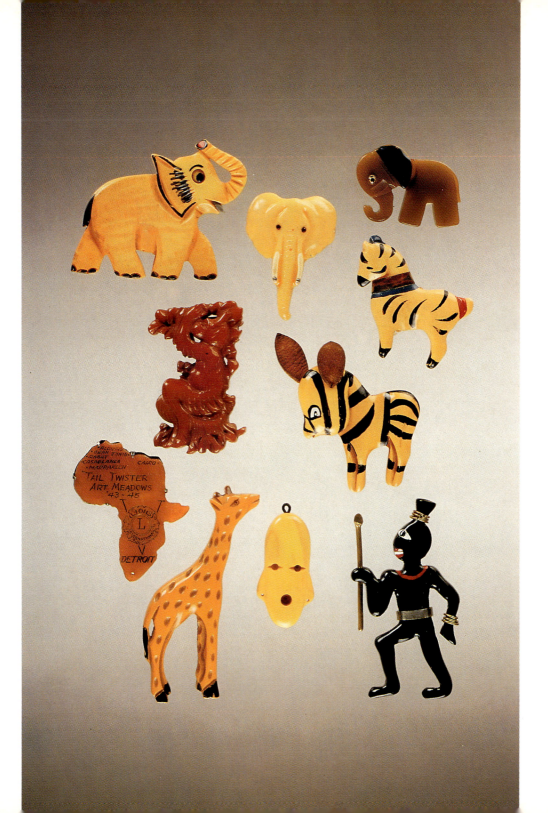

Out of Africa.

Horsing around.

Pegasus

Horse head brooch—sterling mane—rare.

Feathery Floridians.

Yellow bird brooch.

Birds of a feather—the top two are trimmed in Lucite.

Pelican.

Creatures and critters.

Bakelite birds.

Water Creatures

Gone fishing.

Seal, otter, and some strange-looking fish, all Bakelite.

Deep sea swimmers.

Critters.

Exotic fish.

Matching crabs.

Wonders of the sea.

Lucite & Celluloid

Celluloid pierced arm band and pair of bracelets. Pair of horse brasses with laminated frames and Betty Boop cheering. German Bakelite necklace. Lucite purse, with hinged Bakelite bracelet inside.

Clear undercarved Bakelite—Lucite purse.

Metal on Bakelite Oriental hat and animal brooches, apple juice clear pendant and bracelets—Bakelite perfume bottle tops—Lucite purse.

Lucite with sparkles in shoe and evening bag—
Bakelite bracelets and brooches.

Creatures from the sea in Lucite: brooches and a clutch purse.

Glittering Lucite handbag and jewelry.

Bakelite and beaded purse with Lucite jewelry.

Wonderful Celluloid bangles with stones, Deco brooch and mirror—Lucite purse.

Painted and undercut Lucite.

Lucite is an important element in these figural brooches

Lucite molded into glasses frames, bangle bracelets with imbedded materials, and a clutch purse.

Lucite and glitter purse with rhinestone-studded Lucite bangle bracelets and earrings. Oo-La-La.

Purple passion in Lucite and Bakelite—with Lucite purses. Wide Lucite bangle by Chanel, Paris.

"Jelly bellies" set in sterling made by Trifari, center is transparent Lucite. The orchid is a trembler.

Exquisite Lucite hand and orchid trembler
brooch.

Marvelous cat and fish bowl brooch—sterling
with a Lucite center.

Values Guide

A values guide is truly just a guide; prices fluctuate for many reasons—in which part of the country you do your buying, from whom you buy, whether you buy from the best dealer or at an estate sale. All these factors affect the price you pay. Color plays an important part, too: red and black are always the most sought-after colors. Good condition, quality and design in all pieces also help to dictate prices. Consider width, carving, polka dots and multi-colors in bangles. Some pieces are unique and expensive. Buy from a reliable source and buy pieces that are pristine; you will enjoy them for years to come.

Page	Item	Values ($U.S.)
9	Five-color cuff bracelets	2,100–3,400
	Five-color buckle	300
11	Cuff bracelets w/metal & matching pin sets	500–900
	Carved cuff and pin	350–700
	Red & black stretch bracelet, & earrings	250-450
12	Two color bangle	250-900
	Geometric earrings	35-60
14	Polka dot bangles	450–1,100
	Two-color carved bangle, 1/2" to 1"	275–525
15	Bowtie bangles	900–1,400
	Gum drop bangles	350–725
16	Cuff and pin set with metal	500–900
17	Bangle with metal	450–850
	Two-color striped cuff	325–550
18	One-color bangles:	
	1/2" wide	225–325
	1" wide	350–500
	2" wide	400–650
	2 1/2" wide	650–900
19	Stretch bracelet, four-color	350–550
	Link bracelet, four-color	450–700
	Brooch, four-color	275–500
20	Two-color striped cuff	325–550
	Two-color striped brooch	125–250
	Clear, carved, stretch bracelet, floral	400–950
	Clear carved bangle, geometric	400–1,200
	Clear, carved clips, all designs	150–250
	Clear, carved brooch, geometric	275–500
21	Clear carved bangle, floral	350–1,100
	Clear, carved bangle, geometric	400–950
	Clear, carved bangle, fish	950–2,200
	Clear, carved, hinged, geometric	450–1,200
	Clear, carved, hinged, floral	350–1,000
22	Geometric bangles, stripes and dots	400–1,200
	Geometric cuff, four-color	475–850
23	Geometric hinged bracelet with metal	500–1,200